James Douglas, And others

The Hunt and Douglas Process for Extracting Copper from its

Ores.

With an Appendix Including Notes on the Treatment of Silver and Gold

Ores

James Douglas, And others

The Hunt and Douglas Process for Extracting Copper from its Ores.
With an Appendix Including Notes on the Treatment of Silver and Gold Ores

ISBN/EAN: 9783337106164

Printed in Europe, USA, Canada, Australia, Japan

Cover: Foto ©berggeist007 / pixelio.de

More available books at **www.hansebooks.com**

THE

HUNT AND DOUGLAS PROCESS

FOR

EXTRACTING COPPER FROM ITS ORES.

———

WITH AN APPENDIX

INCLUDING

NOTES ON THE TREATMENT OF SILVER AND GOLD ORES,

AND A PLATE.

———

BOSTON:
PRESS OF A. A. KINGMAN.
1876.

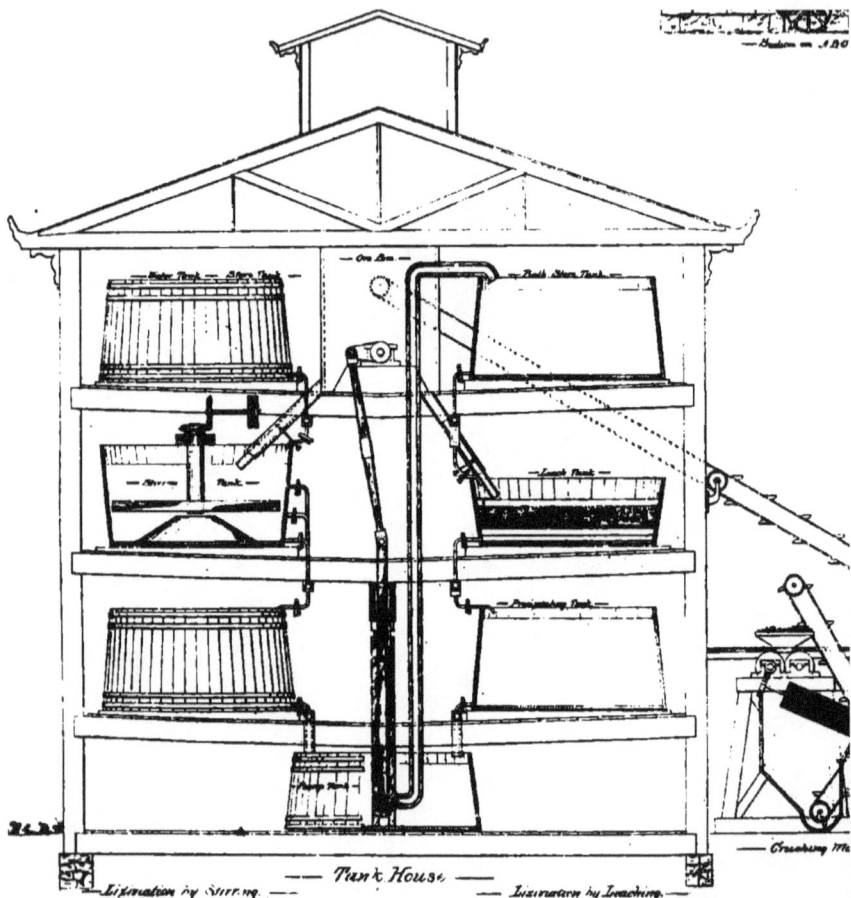

Tank House

Lixiviation by Stirring. Lixiviation by Leaching.

PLAN OF WORKS

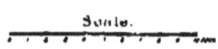

HUNT & DOUGLAS

Scale.

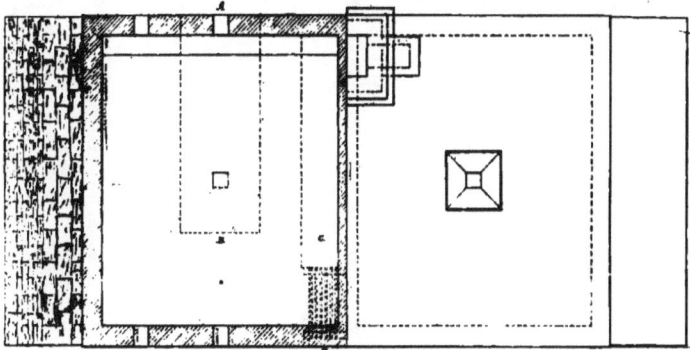

— Half Section Half Plan —
— of —
— Furnace —

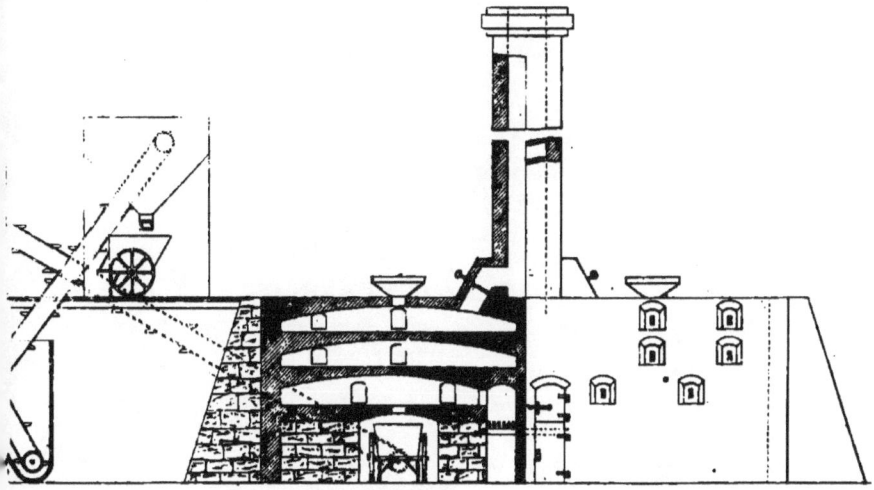

— Cross Section, —
— Calcining Furnace —

— Front View, —
— Calcining Furnace —

TREATING ORES

OPPER PROCESS

TABLE OF CONTENTS.

HUNT AND DOUGLAS COPPER PROCESS.

This is what is technically called a wet method, because the copper is removed from its ores in a dissolved state, the solvent employed in the present process being a watery solution of neutral protochlorid of iron· and common salt. Most oxydized compounds of copper — whether obtained artificially by roasting sulphuretted ores, or found in nature in the forms of carbonates and oxyds, — when digested with such a solution are converted into a mixture of protochlorid and dichlorid of copper, which are dissolved, while the iron of the solvent separates in the form of insoluble hydrous peroxyd of iron. When the solution of the chlorids of copper thus obtained is brought in contact with metallic iron the copper is separated in a metallic crystalline state, while the iron passes into solution, reproducing the protochlorid of iron; thus restoring its solvent powers to the liquid, which we shall call "the bath," and fitting it for the treatment of a fresh portion of copper ore. This process of solution and precipitation can, under proper conditions, be repeated in efinitely with the same bath, the only reagent consumed being the metallic iron.

The chief advantage which wet processes possess over smelting lies in the economy of fuel. To extract copper from a low grade ore by smelting, five or six furnace-operations are necessary, and about one ton of coal is consumed for each ton of ore treated ; while for the various wet processes a single calcination, in which not more than three hundred weight of coal is consumed for each ton of ore, is the only furnace-operation required to obtain the metallic copper in a precipitated form known as *cement copper*. An important item of cost in wet processes is the metallic iron employed to separate the metallic copper from its solutions. The same amount of iron is required to precipitate a ton of copper whether extracted from a poor or a rich ore, but as for the smelting of the latter much less fuel is required, it follows that rich ores are generally treated by smelting rather than in the wet way, any saving of fuel in the latter being more than compensated for by the cost of iron. No general rule however can be laid down to determine what grade of ore can be more profitably treated by one method or the other, inasmuch as circumstances of locality, affecting the cost of fuel and the price of iron, must in each case be taken into account.

The various other wet methods of copper-extraction may be divided into two classes: those in which the previously oxydized ore is treated with hydrochloric or sulphuric acid to dissolve the oxyd of copper, and those in which sulphuretted ore, generally after a preliminary roasting, is calcined with an admixture of sea-salt or of sulphate of soda, by which the copper is converted into chlorid or into sulphate. All of these methods, when properly applied, effect a pretty thorough extraction of the copper, but the cost of the reagents which have to be added to every charge of ore, preclude altogether the use of some of these methods, except in certain favored localities, and render them in almost all cases, it is believed, less economical than the present one with the Hunt and Douglas bath, for which the following advantages are claimed:

I. It is a general method adapted to all compounds of copper, while that by calcination with salt is only applicable to sulphuretted ores.

II. It does not require the addition of reagents such as acids, salt or sulphate of soda to each charge of ore, since in the regular course of the operation the solvent required for the treatment of the ore is constantly reproduced.

III. The bath employed being neutral, certain impurities of the ore, such as arsenic, which pass into solution and contaminate the product in the wet processes, remain undissolved, so that a purer copper is obtained.

IV. As the solution obtained is neutral and free from persalts of iron, there is no unnecessary waste or consumption of metallic iron in the process of precipitation. Moreover, as the result of the action of the protochlorid of iron of the bath on protoxyd of copper, one-third of the copper is obtained as protochlorid, and two-thirds as dichlorid. Now since the latter requires for each one hundred parts of copper precipitated only forty-five parts of iron, it is found in practice that not more than three-quarters of a ton of iron are consumed to precipitate one ton of metallic copper, while in the other methods, in which the copper is obtained as protochlorid, the consumption of iron amounts to a ton, and in many cases greatly exceeds it.

WORKING DIRECTIONS.

The application of the Hunt and Douglas process to the treatment of copper ore may be considered under the following heads:

I. Grinding the ore.
II. Calcining the ore.
III. Dissolving the copper.
IV. Precipitating the copper.
V. Melting and refining the copper.
VI. Arrangement of the plant.

I. *Grinding the Ore.* The degree of fineness to which the ore must be ground will depend entirely upon the character of the gangue. If the metal be scattered in fine particles through an impermeable rock, it will be necessary to grind it to the size of sand, so that the copper, if a sulphuret, may be exposed to the oxidizing action of the air during calcination, and to the solvent action of the protochlorid of iron bath during lixiviation.

If, on the contrary, the copper-sulphuret be mixed, as is often the case, with iron pyrites, which by calcination becomes porous, the ore need not be ground so fine. Experiment in each case must determine the point, and upon the decision must depend the machinery which should be chosen to effect the grinding;— Cornish rolls being preferable for coarse crushing and stamps for finer work. Two pairs of rolls,—one pair of 24 or 30 inches diameter, and one pair of 12 or 15 inches, with a screen between them to sift out what is not broken sufficiently fine by the upper pair, will crush about twenty tons of stuff in twenty-four hours so that it will pass through a sieve of fifteen holes to the linear inch, a degree of fineness sufficient for most ores. A rock-breaker with jaws set close may be substituted for the upper pair of rolls.

II. *Calcining the Ore.* It is not necessary to calcine carbonates or protoxyds, but mixtures in which there is a large proportion of red or dinoxyd need a slight roasting to convert at least a part of this into protoxyd ; while all sulphuretted ores require much more calcination. The mode of effecting this will vary with the character of the ore. When it contains 20 p. c. or upward of sulphur, it may be broken into lumps of an inch or more in diameter, and exposed to a preliminary roasting in heaps or kilns, whereby, without the aid of fuel, the greater part of the sulphur will be driven off, and the metallic ingredients more or less completely oxydized. The lumps thus partially roasted should then be crushed and calcined in a muffle or reverberatory furnace. The calcination of all ores in an earthy gangue must be effected wholly in such furnaces.

The first rule in roasting is to expose the ore at the beginning to a low heat, which is to be gradually increased as the sulphur is driven off. If the temperature be too high at the commencement of the operation, the ore, if highly sulphuretted, may become softened and agglutinated or fritted, after which it is impossible to effect a proper roast. But even if this should not happen, too high a heat at first, or indeed at any stage of the process, brings the copper into a condition in which it is difficultly soluble in the bath. A long furnace is more easily managed than a short one, since in the former the fire can always be kept strong and the ore moved forward from a cooler to a hotter portion, while in a short furnace the gradation of heat can only be attained by close attention to the firing.

A long muffle furnace always gives a good roast, as the tile floor protects the ore from excessive heat, and there is sure to be an oxydizing atmosphere in the furnace, which is not always the case in a reverberatory, where the flame comes in contact with the ore. But the construction of the muffle furnace is expensive, and a cheap and efficient furnace is a three-hearth reverberatory. When a number of such furnaces are needed, they may be built side by side, in a row, the rabbling-doors opening before and behind, and the arches of the whole row being supported by a stone buttress at each end, — the only binding necessary. The fire-boxes of adjacent furnaces are placed side by side. The dimensions which have been found advantageous for these furnaces are as follows : lower hearth ten feet wide by sixteen feet long; upper hearths twelve feet wide by fifteen feet long. The lower hearth is contracted in width by the fire-place, and the upper hearths in length by the flues which lead from hearth to hearth. The details of construction are shown in the accompanying plan.

The advantages of such a form of furnace are cheapness of construction and economy of heat, on account of the exposure of a less amount of cooling surface than in the long reverberatory with rabbling-doors on the side. On the other hand the upper hearths are not very accessible to the rabblers. If such a furnace be used, the heat should only be sufficient to thoroughly dry and warm the ore on the uppermost hearth. Oxydation should take place, with the elimination of the greater part of the sulphur, on the second hearth, so that when the ore is exposed to the higher temperature of the lower hearth there may be no danger of fritting. The quantity of ore which may be roasted in such a furnace will depend on the character of the ore and the proportion of sulphate of copper which it may be desirable to obtain. If the ore is highly sulphuretted and has not received a preliminary roast before grinding, only two or three tons can be calcined in twenty-four hours, whereas double that quantity may be treated if the ore be poor in sulphur. An ore with from 15 p. c. to 20 p. c. of sulphur may be added in charges of 2500 lbs. and shifted from hearth to hearth every eight hours, while one containing from 5 p. c. to 7 p. c. of sulphur may be shifted every five hours.

If the ore contains no carbonate of lime or magnesia (which will deprive the bath of the chlorid of iron in the subsequent operation of solution), the roast need not contain over one-fourth of its copper in the state of sulphate. This will be more than sufficient to repair unavoidable losses in the iron-chlorid of the bath. The presence of portions of these obnoxious elements may, however, make it desirable to obtain in the roast a larger proportion of sulphate of copper (which is soluble in water and by its precipitation by metallic iron yields an iron-salt). To obtain this the ore should be roasted more slowly and in larger charges, say of 5000 lbs. each, in which case the yield of ore from the furnace will be somewhat diminished.

The quantity of fuel consumed will vary with the different ores, but as a rule one cord of wood will suffice for three tons, and one ton of coal for eight tons of ore.

When a sulphuretted ore has been properly roasted it loses, when being rabbled, that apparent fluidity which ore still giving off sulphurous acid exhibits, and when withdrawn and cooled should have a bright red color. If the heat has been too great the color of the cooled ore will vary through dull red to black. There is more danger of having too much than too little heat in the furnace. The ore on the upper hearth should never be in a glow, and that on the lower hearth should never attain a higher heat than dull redness. Besides regulating the heat, it is important to attend to the admission of air. As the roasting of the ore is an oxydizing process an abundance of air is essential to the operation, and that this may be supplied, the furnace must possess a good draft and be provided with openings sufficiently large and numerous. If the furnace be defective in these points the ore will be scorched and its copper rendered insoluble by a reducing action on the lower hearth, while the upper hearth will be liable, at the same time, to become too hot. The more completely the sulphuret of copper is oxydized in the roasting, the more thorough will be the subsequent extraction of the copper, but to oxydize the last traces of sulphuret requires a disproportionate expenditure of time, labor and fuel. Upon the relative value of the raw ore, and of labor and fuel, will therefore depend the degree of thoroughness to which it may be profitable to carry the extraction of the copper at any given reduction-works. While it is desirable to oxydize, as completely as consistent with economy the sulphurets of the ore, it should be borne in mind that a *dead roast*, as it is called, or the elimination of that portion of sulphur which, after oxydation, remains combined as sulphate of copper, is to be avoided, since, as already pointed out, to provide for unavoidable loss of chlorid of iron it is desirable to leave a portion of sulphate of copper in the roasted ore. The composition of the roast may be seen from the following examples.

At the Ore Knob Mine in North Carolina, the average of the ore roasted by this process was, according to Mr. Olcott, (Trans. Amer. Inst. Min. Engineers, vol. III. p. 395.):

Copper as sulphate . . . . . . ,	3.76
Copper as oxyd	7.75
Copper as sulphid	.39
	11.90

At Phœnixville, Pennsylvania, where the ore contains a considerable quantity of carbonate of magnesia, the effect of which has to be neutralized by a large proportion of sulphate of copper, and where charges of 5000 lbs. of ore are calcined for twenty-four hours on each hearth of the dimensions above given, the roasted ore has the following average composition :

Copper as sulphate	1.25
Copper as oxyd	1.10
Copper as sulphid	.40
	2.75

For the method of determining by assay the composition of the roasted ores, see Appendix, p. 21.

III. *Dissolving the Copper.* The solvent or bath employed for the extraction of the copper is, as has been stated, a neutral solution of protochlorid of iron with common salt. This protochlorid may be obtained in various ways. In localities where acids are cheap it is easily made by dissolving scrap iron in diluted muriatic or sulphuric acid; the first yields directly protochlorid, the second protosulphate of iron, which when mixed with a solution of salt gives rise to the protochlorid, together with a portion of sulphate of soda. In places where acids are not so easily had, the commercial protosulphate of iron (green copperas) is the most convenient source of the protochlorid, as explained in the specification. 100 lbs. of the commercial acid and 56 lbs. of scrap iron will make 280 lbs. of copperas. Knowing the relative cost of these substances at any locality, it will be easy to calculate whether it is cheaper to make the copperas or to purchase it. Where highly sulphuretted copper ores or copper pyrites are to be had these, by calcining at a low red heat (as already stated) yield large proportions of sulphate of copper and sulphate of iron, both of which are soluble. By leaching these roasted ores with water and digesting the solution thus obtained with scrap iron the dissolved copper is thrown down as metal, and a solution of protosulphate of iron obtained, which may be mixed with salt to form the bath.

In the original specification of the process it was directed in making the bath by the use of protosulphate of iron to take 280 lbs. of this, (equal to 56 lbs. of metallic iron) and 120 lbs. of salt, sufficient to convert it into protochlorid. These dissolved in 1000 lbs. of water (100 imperial gallons) with a a farther addition of 200 lbs. of salt made the strongest bath, but a weaker one was also recommended in which these same ingredients were to be dissolved in 2000 lbs. of water. Experience has shown that the latter is strong enough for the treatment of all ordinary ores.

The bath may be brought in contact with the ore either by percolation in leaching tanks, or by agitation in vats arranged with stirrers. If the ore be finely ground and slimy, the latter must be used, but if it is coarse, and contains nothing which when wetted will form mud, it is best treated by leaching. When agitation is required the tanks should be round, ten or twelve feet in diameter, and five or six feet high, and made of three-inch staves. A convenient stirring apparatus consists of two oblique blades fixed to the base of a vertical shaft, which rests on the vertex of a conical bottom. The tips of the blades should reach to within an inch of the sides of the tank, and be raised about fifteen inches above the level of the bottom of the tank at the periphery. The object of thus elevating the stirrer on a cone above the bottom is to permit the ore to settle below the blades, so that the stirrer, after having been stopped, can be started at will ; whereas were the bottom flat and the distance between it and the blades the same at all points, the ore would accumulate around the shaft and thus escape agitation. The stirrer should make about twenty revolutions a minute. A vat of the above dimensions, having a capacity of about 3000 gallons, and two-thirds filled with bath, will serve to agitate and dissolve the copper from 3000 lbs. of roasted ore containing five or six p. c. of copper oxyd in six to eight hours, the temperature being from 120° to 150° F. The stirrers are then stopped, the whole allowed to settle, the clear liquor drawn off into the precipitating tanks, and the muddy portions into settling tanks, after which the residue may be washed, first with bath, and then with water, to remove the adherent copper solution.

When percolation can be adopted it is preferable to stirring, since, though the operation is slower, we are enabled to dispense with the settling tanks which the latter plan requires, and the handling of the slimes which accumulate in these. Moreover, as the solution of the copper takes place in the mass of ore out of contact of air, a larger proportion of dichlorid of copper is found and less iron is lost by oxydation than when the solution holding the dissolved iron and copper salts is exposed to the air by constant agitation.

The vats for filtration are made of wood or of brick. For the latter the bricks are laid in Roman cement and coated within by a layer of the same cement mixed with silicate of soda. This, when afterwards washed with a solution of chlorid of calcium, forms a coating which resists the action of the metallic salts of the bath. If wood be used the vats may either be square or round, but in any case they should be somewhat wider at the top than at the bottom, otherwise the settling and contraction of the moistened ore will leave a space along the walls through which the bath may descend without percolating the mass. Filtering vats need not be more than three feet high. The filter may be made by laying on the bottom of the vat three inches of small stones, broken cinder or coke, and covering this by a layer of coarse sand upon which the ore may be laid to the depth of one or two feet according to its coarseness or fineness. Instead of this arrangement a false bottom, consisting of perforated planks or of narrow boards loosely laid ogether, may be covered over with coarse sacking upon which the ore is spread. A hole in the side near the bottom of the tank, into which is fitted an india-rubber tube provided with a squeezer or pinch-cock, gives vent to the liquor after its passage through the ore. The vats should be fitted with close covers so as to exclude the air and retain the heat; these are provided with a small hole through which enters a tube to supply the bath. It is well to spread the ore in the vats already partially filled with heated bath, as when thus wet down it will not cake but will permit the bath to percolate uniformly through it. When the desired quantity of ore has been added a wooden float should be secured beneath the opening in the cover so that the bath, as it flows in, may fall thereon, otherwise it would make a depression in the mass and thus the percolation would be unequal. About two or three inches of bath should be kept on the top of the ore, and it should be supplied as rapidly as it escapes from the tube below. When the outflowing liquid is found, by testing with a bit of iron wire, to contain no more copper than the liquid entering above, we know that the soluble copper has been removed and it only remains to stop the supply of bath, allow the layer above to filter through and then displace that which remains in the pores of the exhausted mass by the addition of a little water. The extraction of the copper by filtration may not be completed in less than three or four days, the time of course depending on the richness of the ore and the strength and the temperature of the bath.

The solution of the copper is much accelerated by heating the bath, which may be done by the injection of steam. If the liquid be heated to from 120° to 180° F. it will flow through the ore in the leaching tanks with a very little reduction of temperature, and the heat generated in the process of precipitation will, if the tanks for this operation are well covered, maintain the bath in these at a sufficient temperature to ensure a quick separation of the copper, so that it is only the liquid in the store tanks that will require heating by steam.

The bath after it is withdrawn from the precipitating tanks generally contains a little copper. If, however, care be taken to leave it there till the whole of the copper is separated, the liquid will then be without action on metallic iron, and steam coils may be used to heat it in the store tank, or in passing from this to the leach vats it may be made to pass through a coil of iron pipe heated by a stove.

Where kilns are used for roasting, the heating of the liquors, as well as the evaporation of the excess of liquid derived from the wash-waters, may be effected in Gay-Lussac towers, which are small brick or stone chambers, tall and narrow, filled with fragments of coke or broken bricks, in which an ascending current of the hot air and sulphurous vapors meets a descending current of the liquid. The hot gases from the kiln or from a muffle may also be utilized by drawing them through a pipe from four to six inches in diameter by means of a small steam jet introduced at the bend in an injection pipe, which at that point should be contracted to two inches, and may dip two feet or more into the liquid in the store tank. Such an arrangement saves steam and serves to impregnate the bath with sulphurous acid, which in its passage through the ore in the filtering vats serves to attack the separated peroxyd of iron, converting it into soluble protosalt.

The use of the sulphurous acid fumes, which thus serve to supply the losses of protochlorid of iron, need not be resorted to except in treating native carbonates or oxyds of copper, or such ores as contain carbonates of lime or magnesia or oxyd of lead or of zinc, all of which cause a loss of the protochlorid of iron. In such cases the best mode of applying the sulphurous acid is by using stirring tanks and passing the gas over the surface of the liquid, which is agitated during the solution of the copper. The gas should be as little diluted with air as possible. If the roasting kiln or muffle furnace be connected with the stirring tank by an earthenware tube which enters either the cover or the side of the tank at a point opposite to that by which a wooden tube (best connected with a flue) gives exit to the unconsumed gas, a sufficiently rapid current of the gas will be kept up, and will be readily absorbed by the liquid in the tank.

It is seldom, however, that the process is thus complicated by the necessity of using sulphurous acid gas, for unless the objectionable matters mentioned above are present in considerable quantities in the ore, this, if a sulphuret of copper, will yield by careful calcination, as already explained, enough sulphate of copper to compensate for the loss of chlorid of iron which these would occasion.

By the introduction of steam for heating and of water for washing the residue, the volume of the bath becomes slowly augmented. To reduce this it is therefore necessary to resort to evaporation, and for this purpose it has been found that in the case of the three-heated reverberatory furnaces already described, the upper hearth may with advantage be used for this purpose. A shallow basin six inches deep, of which the upper hearth is the bottom, is built of brick lined with Roman cement prepared as already described for brick leaching tanks, with silicate of soda and chlorid of calcium. Through the middle of it passes a funnel or hopper connecting with the second hearth and opening on the roof of the furnace for the purpose of charging the ore. The roof over the pan should be higher than if it were over a calcining hearth, in order to give ample room to use a scraper with which to remove the crystals of salt which separate during evaporation, and two doors instead of three should open at each end so as to give free access to the whole surface of the pan. Care should be taken to keep the floor of the pan free from accumulations of salt, else a crust will be formed which is difficult to remove. When this arrangement is used, the liquors from the store tank are run into the pan where they are exposed to the escaping current of hot air and gases by which they are rapidly evaporated, while at the same time they absorb a considerable amount of sulphuric acid (which, together with sulphurous acid, is formed in the slow roasting of pyritous ores), and thus become strongly acid. An evaporator of this kind, with a surface of 100 square feet, is easily and cheaply constructed, is tight and durable, and will evaporate a layer of four inches of liquid in twenty-four hours, by the waste heat. If the three hearths are required for calcination, large shallow tanks of the kind described may be constructed between the furnaces and the stack, so that the whole current of the hot gases shall pass over the surface of the liquid, thus dispensing with the use of steam for heating the liquors and evaporating the bath at the same time.

IV. *Precipitating the Copper.* The copper liquors, whether taken from the stirring or settling tanks, or flowing from the leaching-vats, are received in tanks of any convenient size, where in contact with metallic iron the chlorids of copper are decomposed, and the copper is precipitated in crystalline grains, plates or crusts, the texture of which will vary according to the strength of liquors. Wrought-iron precipitates the copper more rapidly than cast iron, but where this latter is the cheaper it should be used. If small scrap is employed it must be spread

2

on trays arranged in the vats. In the Appendix will be found described a method of cleaning tinned iron scrap by which the tin may be saved and the iron fitted for precipitating copper. If the residue after the extraction of the copper is a nearly pure oxyd-of iron, this may be reduced to a spongy metallic iron by heating it for some hours at a red heat with pulverized coal in a closed vessel. This spongy iron, which may also be easily made from ordinary iron ores, precipitates copper very rapidly from its solutions, and is used for that purpose in England, where it is prepared in a reverberatory furnace. Details of the preparation and use of the iron sponge will be found in the Appendix.

From the precipitating vats the liquor which has been deprived of its copper, and in this process has been recharged with protochlorid of iron, is drawn off after twenty-four hours or more, and pumped up into the store tank, which should be at a higher level than the leach vats. It is then ready for the treatment of fresh portions of ore. A working drawing of a cheap pump made entirely of wood, is included in the plan annexed.

V. *Melting and Refining the Copper.* The precipitating tanks are emptied from time to time, the cement copper is washed with water and if small iron scrap has been used, is passed through a screen or sieve to separate any fragments of the latter. It is then dried at a gentle heat, when it is ready for refining. In the treatment of copper obtained in wet processes it is customary to melt the cement with a portion of matte or sulphuretted ore, and thus obtain a crude copper which is refined by a second fusion. Experience on a large scale has, . however, shown that the purity of the cement obtained by the present process is such that but a single fusion is required to convert it into fine copper. The dried, or even the moist cement, is melted down in a furnace such as is used for refining blister copper, poled in the usual manner and then cast into ingots. It is found advantageous to mix the cement with one or two hundredths of coal dust, and if compressed into blocks it can be handled with greater advantage than if in a loose powder.

VI. *Arrangement of Plant.* It is well, when it can be done, to choose a hill-side as the site for works for carrying out this process, so as to place the leaching vats below the level of the calcining furnaces. Above these vats should be a water tank and also a store tank for bath, from which it can be made to flow into the vats placed in rows on the lower level. Below these should be placed the precipitating tanks and still lower a large tank into which the bath when deprived of copper can be allowed to flow, and from which by means of the wooden pump it is to be pumped up into the store tank for redistribution, thus establishing a continuous circulation. Wooden tubes securely coupled together are the best conductors for the bath. A horizontal line of such tubing should run above the leaching vats and be connected with the store tank by a piece of india-rubber tube or hose, which can be closed at will by a wooden squeezer. From this line of wooden tubing the bath is to be conducted to each leach tank by an india-rubber tube, the flow through which is to be regulated by squeezers. From the leach tanks the copper liquors should be conducted through similar india-rubber tubes into a covered trough or launder, running the whole length of the row of precipitating vats. Such a trough is better for this purpose than a closed tube, for the reason that when the bath is too cool or does not hold a sufficient amount of salt to retain the whole of the dichlorid of copper in solution, a portion of this may be deposited and fill up the tube, while the launder can be watched and this state of things guarded against.

In localities where a hill-side cannot be chosen for the site it will be better to place both the leaching and the precipitating vats on the same level with the calciners, for it is easier and cheaper to pump the copper liquors into the precipitating tanks than to elevate the ore. In the plan, however, for clearness of illustration the tanks are shown on successive levels in a building of three stories.

APPENDIX.

I. Assaying for Copper and for Iron.

' The following directions will enable any one, even without a knowledge of chemistry, to make the tests necessary for the successful working of this process; namely, the determination of the quantity of copper in the ore, the character of the roast, and the condition of the bath.

Copper. The most expeditious and convenient method for the determination of copper depends upon the property of cyanid of potassium to decolorize the deep-blue solution which is got by adding ammonia in excess to the protosalts of copper. Begin by dissolving in clear water ordinary commercial cyanid of potassium, in the proportion of about a half a pound to a gallon. Next weigh out carefully five grains of pure metallic copper, such as copper foil; dissolve it in a little nitric acid ; add water to the bulk of about four ounces, and then caustic ammonia (*liquor ammoniæ*), until a deep-blue liquid is obtained. Next fill a graduated tube, known as a burette and generally divided into cubic centimeters and fractions thereof, with the solution of cyanid of potassium, and allow this to drop from the burette into the copper solution, till the blue color disappears, first giving place to a pinkish hue. Great care must be taken to add slowly towards the end of the operation, so as to avoid the addition of an excess of the cyanid. Now read off on the graduated tube the quantity of the solution of cyanid which has been required to produce the decoloration of the solution of five grains of copper. Suppose 25.0 cubic centimeters (c.c.) have been consumed; then it is clear that one grain of copper corresponds to 5 c.c. of the cyanid solution.

To make an assay of a copper ore, reduce it to fine powder, and weigh out carefully 25, 50, or 100 grains, according as it is rich or poor in copper. Add to this, in a small flask or beaker-glass, common nitric acid, sufficient in quantity to cover the ore. Apply heat to dissolve the copper, and if the ore is a sulphuret boil it, until, on adding a fresh portion of acid, no more red vapors are given off. Now add from four to eight ounces of water, and pour in ammonia till the blue color is obtained, and the liquid shows by its

(20)

smell that a slight excess of ammonia has been added. By this means the dissolved iron (which is almost always present in the ore) will be separated as a bulky, reddish-brown peroxyd, which rapidly settles, leaving the clear blue liquid above. Add to the mixture at once the cyanid of potassium from the burette, stirring the while, and allow the suspended·oxyd of iron to settle from time to time, so as to judge of the progress of the operation from the color of the clear liquid, which soon appears above the subsiding precipitate of the brown peroxyd. The ̦operation may require ten or fifteen minutes. When the color has faded out, as in the previous example, note on the burette the quantity of the cyanid solution consumed. Suppose this to be 48 c.c. Now as 5 c.c. are equal to a grain of copper, we have the proportion 5 : 48 :: 1 : 9.6; so that, if the quantity of ore was 100 grains, the ore contains 9.6 p. c. of copper, or, if 50 grains were employed, 19.2 p. c.

In testing the roasted sulphuret ore, there should be determined in each sample three things : (1) the quantity of sulphate of copper, a portion of which, as already explained, should always be present; (2) the amount of oxyd of copper; and (3) the quantity of unoxydized copper, existing in the form of sulphuret. As sulphate of copper is readily soluble in water, it is only necessary for the first determination to boil a weighed portion of the ore with a little water, pour off the clear solution, wash the residue with cold water, add ammonia to the liquid, and proceed as before. Muriatic acid diluted with twenty-five times its bulk of water will, at a boiling heat, readily dissolve the oxyd of copper from the roasted ore, without attacking the sulphuret. If then we boil the roast for two or three minutes with a sufficient quantity of such dilute acid, allow the undissolved portion to settle, and wash it thoroughly with several waters, the only copper left in the residuewill be that of the sulphuret which has escaped roasting and which may be dissolved by boiling with nitric acid.

Suppose 100 grains of a roasted ore to be boiled with water: the clear solution is poured off, mixed with ammonia, and treated with the standard solution of cyanid of potassium, of which 15 c.c. are consumed. Then 5 : 15 :: 1 : 3; so that the sample holds 3.0 p. c. of copper as soluble sulphate. The residue is now boiled for two or three minutes with two ounces or more of dilute muriatic acid, as already described, and the solution then obtained decanted and boiled for a minute with a few drops of nitric acid, to convert any dichlorid into protochlorid of copper (this is necessary to get the full color, because the ammoniacal solution of the dichlorid is colorless). Ammonia is now added, and then the standard cyanid solution. If of this 32 c.c. are required to decolorize, we have 5 : 32 :: 1 : 6.4, or 6.4 p. c. of copper in the state of soluble oxyd. The insoluble residue is then boiled with nitric acid till red fumes are no longer given off, which may take as much as five or ten minutes, water is added, and excess of ammonia, when it is found that 2 c.c. of the standard solution of cyanid are required for decolorizing. We have thus 5 : 2 :: 1 : 0.4; so that there remained 0.4 p.c. of copper as insoluble sulphuret. The result will then stand as follows : —

Copper as sulphate 3.0
"   " oxyd 6.4
"   " sulphid 0.4
                                                    ———
                                                    9.8

In thus assaying a sample of ore, the percentage of copper in which is known, we may dispense with directly determing the oxyd, and by deducting from the total the amount of soluble sulphate and insoluble sulphuret, we find the quantity of oxyd by loss. In the regular working of the process, in fact, it is only necessary to determine the sulphate and the sulphuret of the roast. The former is important, because as already explained, it is necessary to have a certain portion of sulphate to make up for the loss of protochlorid of iron in the bath, and the second because the copper which remains unroasted in the form of insoluble sulphuret is not extracted by the bath, and therefore is lost.

The above process of copper assaying, though the most simple and expeditious, is not directly applicable to ores containing silver, zinc, cobalt, nickel or manganese. The absence of the first may always be ensured by adding a few drops of muriatic acid to the solution, by which silver is thrown down as a white insoluble chlorid. When, as is not infrequently the case, one or more of the other metals mentioned are present, the following modification of the process may be adopted. The solution from the roasted ore by water, mixed with a little muriatic acid, or the solution of the oxyd of copper in this acid, is digested with metallic iron or zinc, by which the whole of copper is thrown down in a spongy metallic form, while the metals zinc, cobalt, nickel and manganese, which would vitiate the operation of decoloration, remain dissolved. Iron wire or a slip of sheet zinc may be used, and if heat be applied the precipitation of the copper will generally be finished in half an hour. It may be known to be complete when a bit of clean bright iron dipped for a minute in the liquid gets no color of copper upon its surface. The metallic copper precipitated is carefully separated from the iron or zinc, washed with water, dissolved in a little nitric acid, mixed with an excess of ammonia and determined by the use of a cyanid solution as before.

When a copper ore has been dissolved by nitric acid it is necessary to get rid of this acid before precipitating the copper. To this end the nitric solution is to be evaporated nearly to dryness, then mixed with about as much muriatic acid as had been used of nitric, and again evaporated nearly to dryness. Water is now added, and from the solution the metallic cop per precipitated by iron or by zinc, and determined in the manner already described. The amount of copper held in solution at any time by the bath, is also readily determined by precipitation with iron or zinc. It is well to add thereto a few drops of muriatic acid, and heat quickens the process. If the quantity of copper precipitated is considerable, we may conveniently wash it with water several times, then with a little alcohol, dry it at a gentle heat and weigh it directly. Smaller quantities are however best determined by dissolving the precipitated copper in nitric acid, adding ammonia and the standard solution of cyanid.

The cyanid solution is liable to slow decomposition, by which its strength is diminished, so that from time to time the stock of the solution (which should be kept in well-stoppered bottles and out of the direct light,), must be standardized afresh by a solution of a known quantity of copper as already explained. If the solution becomes turbid it may be filtered through paper.

Iron. The following simple and rapid mode of assaying for iron in the bath, is based upon the fact that the deep crimson color of a solution of permanganate of potash (mineral chameleon) is at once destroyed by a protosalt of iron, such as protosulphate or protochlorid. It is therefore only necessary to add to the acid and dilute solution of a protosalt of iron, which is colorless, a solution of the crimson permanganate of known strength until the tint of the latter remains in the liquid. We begin by dissolving five grains of pure clean iron wire (like that used for piano strings) in dilute sulphuric or muriatic acid, and then add half a pint or more of pure cold water, and a few drops of sulphuric acid. Now pour from a tube or burette graduated like that used for the cyanid solution in copper testing, a dilute solution of the permanganate, rapidly at first, and then more slowly constantly stirring, till the iron solution acquires a faint crimson tint. The quantity of chameleon solution employed, divided by the number of grains of iron, will give the number of cubic centimeters which correspond to a grain of this metal. As the bath used in the Hunt and Douglas process contains the iron in the state of protochlorid it is only necessary to take a known volume of it and proceed as above to determine its content of iron. If, for example, the permanganate solution is of such a strength that a solution of five grains of iron requires 42 cubic centimeters of chameleon solution to color it, equal to 8.4 c.c. for a grain of iron, we take two fluid drachms (a quarter of a fluid ounce) of the bath, add thereto twenty drops of sulphuric acid and half a pint of pure cold water and find that it requires 12.6 c.c. of chameleon to give it a crimson tint, then 8.4 : 12.6 :: 1.0 : 1.5 so that this quantity of the bath contains 1.5 grains of metallic iron, equal to 6 grains to the fluid ounce. By testing in this manner from day to day a measured quantity of the bath, its loss or gain in iron can be very easily determined.

The solution of the permanganate should be kept in a glass-stoppered bottle and should be tested from time to time with a solution containing a known weight of iron. For this purpose, instead of iron wire, which requires some time to dissolve in acid, we may conveniently use the double sulphate of protoxyd of iron and ammonia (ammonio-ferrous sulphate). The green readily soluble crystals of this salt contain exactly one-seventh their weight of iron, so that 35 grains of this salt, with 20 drops of sulphuric acid dissolved in half a pint of water, correspond to a solution of 5 grains of metallic iron.

II. Preparation and Use of Iron Sponge.

The use of spongy metallic iron for precipitating copper in a metallic condition from its solutions was proposed in 1859 by Mr. William Gossage, of Widnes, England, to whom letters patent for improvements in extracting copper were then granted in Great Britain, the dates of the specification being March 7 and September 7 of that year (patent No. 194). He claimed the extraction of copper from the residues of calcined pyrites by the use of solutions of persalts of iron. The copper thus dissolved being thrown down by metallic iron, he obtained a protosalt of iron (protosulphate or protochlorid), which he converted by evaporation and exposure to the air at a heat below redness into a mixture of insoluble peroxyd and of a persalt soluble in water, which could be used to dissolve another portion of copper. In the case of silver-bearing ores he combined the persalt of iron with common salt, to dissolve the chlorid of silver formed.

In the same patent he claimed for the precipitation of the dissolved copper finely divided or spongy iron prepared from calcined pyrites residues containing a portion of copper, " by mixing them with about one-fourth their weight of coal, coke, or charcoal, in coarse powder, and introducing the mixture into an oven or closed furnace kept at a red heat for about twelve hours, or until the greater part of the oxyd of iron contained therein has become converted into metallic iron. I then withdraw the product as quickly as possible, and I receive it either in vessels containing water, or in vessels from which the air can be excluded till the contents have become sufficiently cooled." " I employ metallic iron obtained by the means herein described for the precipitation of copper from its solutions, in the same manner that other metallic iron is employed for effecting such precipitation." In 1862 Gustav Bischof, Jr., obtained letters patent in Great Britain for the manufacture of spongy iron for precipitating copper, the materials used and the mode of working being essentially the same as those of the earlier patent of Gossage, and in 1868 a United States patent for the same invention was granted Mr. Bischof and his assignee Mr. John S. Kidwell.

The manufacture of iron sponge for this purpose as carried on at Newcastle, in England, is described in a paper read before the Newcastle Chemical Society by Mr. Gibbs, and published in Engineering and in Van Nostrand's Engineering Magazine for October, 1875, from which the following notes and extracts are made:

The ore now treated for copper by wet process in England is the residue · of the Spanish pyrites (of which nearly 400,000 tons are there annually consumed), which has been calcined to extract its sulphur, and is then known as *burnt ore.* It contains 3 or 4 p. c. of copper, as much sulphur, a little arsenic, lead, and zinc, and about 2 p. c. of silicious matter. By calcination with about 15 p. c. of salt, in what is known as the Longmaid or Henderson process and washing with water and with dilute muriatic acid, the copper, sulphur and arsenic are removed, and the residue, known as *purple ore*, is nearly pure peroxyd of iron. Both Gossage and Bischof proposed the use of the burnt ore for the manufacture of the sponge on the ground that the copper present would be obtained with the precipitate; but the arsenic which remains in the burnt ore is such an objectionable impurity that, according to Mr. Gibbs, only the purified or purple ore is now employed for the manufacture of sponge for copper precipitation. He thus describes the furnace and its working:

" This is essentially a reverberatory furnace, 30 ft. long, with a provision for conveying the flame under the hearth, after it has passed over the charge. The hearth of the furnace is 23 ft. long and 8 ft. wide, and is divided into three working-beds by bridges. Each bed has two working-doors on one side. The doors slide in grooves, and close air-tight. The fire-box is 4 ft. by 3 ft., with bars 4 ft. 8 in. below the bridge, thus allowing for a considerable depth of burning fuel. The fire-door slides in grooves like the working-doors. The hearth is formed of tiles sustained on brickwork partitions, forming flues through which the flame returns after passing over the hearth. From these flues the flame drops by a vertical flue alongside the fire-bridge, to an underground flue communicating with a chimney. The entrance to the latter flue is provided with a fire-tile damper, which is closed whenever the working or fire-doors of the furnace have to be opened. A cast-iron pan, 20 ft. by 10 ft., is carried by short columns and girders over the furnace-roof. In this pan the ore is dried and mixed with coal, and from it is charged into the hearth through cast-iron pipes , built into the furnace-arch. The furnace is elevated on brick pillars, to allow of iron cases running under it, and it is worked from a platform of cast-iron plates. A vertical pipe, 6 in. diameter, passes through the hearth of the furnace, inside each working-door; and through these pipes the reduced iron is discharged into iron cases placed beneath. These cases are horizontally rectangular, and taper upwards on all sides. The cover is

3

fixed, and in its centre is a hole 6 in. diameter, with a flange upwards which serves to connect the case with the discharging-pipe. The bottom of the case is closed by a folding-door, hinged on one side, and secured by bolts and cutters on the other. The case is fitted with four wheels, clear of the door, and is covered with a cast-iron plate, fitting loosely into the opening on the upper side. It stands 4 ft. 8. in. high, and has a capacity of 12 cubic feet."

" The furnace-hearth being at a bright red heat, each of the three working-beds is charged with 20 cwt. of dry purple ore, and 6 cwt. ground coal from the cast-iron pan under the roof. The fire and working doors are closed, and the only air entering is that through the fire, in working which care is taken to prevent the mass of burning fuel getting hollow. The charge in the first bed from the fire-bridge is reduced in from nine to twelve hours; in the second, in eighteen hours; and in the third, in about twenty-four hours. Each charge is stirred over two or three times during the period of reduction. Before opening any door the flue-damper is closed, to prevent a current of air entering over the charge. On the complete reduction of the charge on any working-bed, two cases are run under the bottom pipe, to which their mouths are luted by clay, and the charge is quickly drawn into them, by rakes worked through the doors. The cases are then closed with cast-iron plates. In about forty-eight hours the iron is cooled sufficiently to be discharged; and this is simply done by raising the case by a crane, and knocking out the cutters fastening the hinged door on the bottom, when, from the tapering form of the case, the mass of reduced iron falls out readily. The sponge is ground to powder under a pair of heavy edge-stones, 6 ft. in diameter, and is passed through a sieve of fifty holes per linear inch."

The heat being a bright red, the reduction of the charges is sure to take place on the hearths in the times specified; but it may be completely though more slowly effected at a very dull red. The material used is said to contain 95 p. c. of peroxyd of iron; and the product holds metallic iron 70.40, peroxyd 8.15, protoxyd 2.40, sulphur 1.07, copper 0.24, with small portions of lead and zinc, 7.60 of carbon, and 9.80 of silicious residue; so that about 90 p. c. of the iron of the charge is reduced to the metallic state. If we suppose that four charges of 20 cwt. each of such ore are reduced in twenty-four hours, the yield would be nearly 48 cwt. of metallic iron.

" In using spongy iron in precipitating copper, the liquids are agitated by an air-blast while the iron is gradually added. By this means a very perfect mixture is obtained, and a copper precipitate can be readily produced, containing not more than 1 p. c. of metallic iron. As compared with precipitation by scrap iron, the economy of space required and facility of manipulation are very great. On the side of spongy-iron precipitation are cheapness of material and economy of application; while against it is the presence with the precipitated copper of the unreduced iron oxides and excess of carbon from the reduction. In employing spongy iron, the copper-extractor has the production of the precipitant in his own hands, and avoids the troublesome handling of a material so cumbrous as scrap iron."

It is comparatively rare that the residue from copper ores treated in the wet way, will yield an oxyd of iron approaching in purity to that obtained from Spanish pyrites; and it is clear that an oxyd containing a large proportion of earthy matter is not fitted for the production of iron sponge for copper precipitation, inasmuch as the whole of this would remain as an impurity in the cement copper. The purer native oxyds of iron, however, such as the magnetic and specular ores, may be advantageously employed for the production of sponge, by grinding them to powder and heating with an admixture of coal, as already described.

The production of pure spongy iron, on a commercial scale, has lately been perfected by Mr. T. S. Blair, of Pittsburg, by means of an ingenious gas-furnace in which the ore in small masses is heated to redness with coarsely powdered charcoal. The iron thus prepared, being free from oxyd and from the earthy matters of the fuel, is fitted for the manufacture of fine steel; but this perfected process is not adapted to powdered ores and residues like those from pyrites. As a matter of economy it remains to be seen whether the somewhat impure iron sponge, which may be very economically prepared from these, is to be preferred to the purer sponge, which is made on a large scale and cheaply in Mr. Blair's furnaces.

III. Use of Tin Plate Scrap.

Tin plate, which is iron coated with tin, may be advantageously used for precipitating copper, when, as in the Hunt and Douglas process, the solutions contain protochlorid of copper, together with a soluble sulphate, as sulphate of soda. A heated solution of this kind very readily removes the tin from the iron, and causes its separation in white flakes of insoluble hydrated peroxyd of tin, with the liberation of a portion of free hydrochloric acid,[1] leaving the iron in a state fit for the precipitation of the copper. In practice the tin plate scrap may be used directly in the vats in place of

---

[1] In this reaction protochlorid of copper and metallic tin yield dichlorid of copper and perchlorid of tin, which last, in solution and in the presence of a sulphate, is broken up into hydrated peroxyd of tin and free hydrochloric acid. The final result of the reaction may be thus represented:

$$4CuCl + Sn + H_2O_2 = 2Cu_2Cl + SnO_2 + 2HCl.$$

scrap iron, in which case the tin oxyd remains mixed with the cement copper, and will be more or less completely washed away in the subsequent stages of the process. Otherwise the tin plate scrap may, by a simple arrangement, be immersed for a few minutes in the hot copper solution till the tin is taken off, and may then be removed to the copper-precipitation tanks. The separated tin oxyd may be collected by subsidence, freed from any adhering metallic copper by washing with a portion of the hot solution containing protochlorid of copper, and when thus purified reduced to the metallic state or used for the manufacture of stannate of soda. Ordinary tin plate carries 3 or 4 p. c. of tin.

As tin plate scrap has, in most places, little or no commercial value, it may often be advantageously employed for the precipitation of copper from its solutions. The process above described was made the subject of letters patent granted to Thomas Sterry Hunt, May 19, 1874, (No. 150,957) for " An improvement in precipitating copper by means of tin scrap," the claim being as follows:

1. " The use and application of tin plate scrap or waste for precipitating copper from its solutions, substantially as above described."

2. " The recovery and utilization of the tin from the tin plate scrap by means of its solution and subsequent precipitation as oxyd of tin in solutions containing protochlorid of copper and a sulphate, substantially as above described."

IV. CHEMISTRY OF THE HUNT AND DOUGLAS PROCESS.

The peculiarity of this method is the use of a solution of *protochlorid of iron* and chlorid of sodium to render soluble the oxydized compounds of copper. In the wet process now generally adopted in Great Britain and mentioned on page 25, where the ores are calcined with common salt, this is in great part decomposed with the formation of sulphate of soda and chlorids of copper, which are, in their turn, decomposed when in solution, by contact with iron, with separation of metallic copper and production of protochlorid of iron.[1]

---

[1] The concentrated liquid obtained by leaching the ores in this process at Widnes in England gave, according to Claudet, for a litre of specific gravity 1.24; sulphate of soda 14.41 grammes; chlorid of sodium 6.39; chlorids of copper and other metals 12.75, containing chlorine 6.61, copper 5.28, zinc 0.68, lead 0.057, iron 0.045, silver 0.004, besides a little gold and small but undetermined quantities of arsenic, antimony and bismuth. Of the copper 0.580 was in the state of dichlorid. The silver extracted from this solution by Claudet's method, with iodine, contains about 1.3 p. c. of gold. *Chemical News*, Vol. XXII, p. 184.

The liquid thus obtained, holding an abundance of protochlorid of iron with a little chlorid of sodium, is found to have but a feeble solvent action upon the oxyd of copper and is accordingly thrown away, polluting the rivers, and thus giving rise to serious difficulties in England. Various attempts have been made to utilize the chlorid of iron in these waste liquors for chloridizing copper. Gossage patented a plan which consisted in evaporating them to dryness and heating the residue in contact with air to low redness, by which means there is obtained a mixture of insoluble peroxyd and soluble *perchlorid of iron* (see page 24). Henderson effects the same result by the action of air on the liquors at ordinary temperatures. He also by decomposing the evaporated waste liquors at a strong red heat in contact with silicious matters, gets perchlorid of iron in vapor with some hydrochloric acid and free chlorine, and dissolves these in a solution of protochlorid of iron, thus getting a solution of *perchlorid of iron*, the solvent action of which on oxyd of copper is well known. (British patent of May, 1865, No. 1255, and United States patent, Dec., 1866, No. 60,514.)

To dispense with these tedious and costly processes and enable liquor, containing *protochlorid of iron* to be directly used for the solution of copper, was much to be desired. It was found that when a solution of protochlorid of iron is brought in contact with either protoxyd or dinoxyd of copper, dichlorid of copper is formed, which, being insoluble in water, soon coats over the oxyd and arrests the chloridizing process. To overcome this difficulty, however, it was only necessary to add a hot and strong solution of common salt in which (as in all other solutions of chlorids) the dichlorid of copper has a considerable degree of solubility. The reactions of the two oxyds of copper with protochlorid of iron are unlike. Three equivalents of the protoxyd, containing 95.25 of copper, when brought in contact with an excess of solution of the protochlorid under the conditions just explained, react with two equivalents of it, containing 56.00 of iron, and yield one equivalent of the protochlorid of copper, which is readily soluble in water, and contains 31.75 of copper, and 35.50 of chlorine, and one equivalent of the insoluble dichlorid, in which the same amount of chlorine is united with twice as much, or 63.50 of copper. When the copper is in the state of the dinoxyd, only one-half as much protochlorid of iron is consumed, and there is formed for the same amount of dichlorid as before, one equivalent, or 31.75 of metallic copper. This would remain undissolved if the dinoxyd alone were treated, but metallic copper in presence of an excess of protochlorid of copper is at once converted into the dichlorid, so that if one-half of the oxydized copper in a mixture treated with an excess of protochlorid of iron, is protoxyd and one-half dinoxyd, the whole of the copper passes into the state of dichlorid.[1] For this reason it is necessary, in submitting dinoxyd ores to this process, either to mix them with a sufficient amount of ores containing protoxyd, or to calcine them slightly in the air, so as to convert one-half or more of the dinoxyd into protoxyd of copper.

In the reaction between the oxyd of copper and protochlorid of iron, the iron of the latter separates from the solution as a reddish brown insoluble precipitate of hydrous peroxyd, which carries with it a small portion of chlorine in the form of an oxychlorid of iron, due to secondary reactions and in part to the action of the air upon the solution of protochlorid of iron. The amount of chlorine thus removed, and consequently lost to the bath, was found, in carefully conducted experiments, to vary from 5 to 10 p. c. of that originally united with the iron, that is to say for 100 parts of protochlorid of iron consumed in chloridizing copper, the regenerated bath will contain from 90 to 95 parts. This loss of chlorine must in all cases be supplied if the strength of the bath is to be kept up, an end which is readily obtained in one or two ways. When sulphuretted ores are oxydized, there is always formed a portion of sulphate of copper, which, with careful roasting (page 10), may equal one-fourth or even one-half of the copper present. This sulphate when decomposed by metallic iron gives protosulphate of iron, which, by its reaction with salt yields, as we have seen in the preparation of the bath, sulphate of soda and protochlorid of iron, which in ordinary cases more than suffices to supply any loss of chlorine. If, as sometimes happens, there is found too large a portion of these compounds, this may be corrected by adding to the bath, *previously freed from copper*, a small quantity of slaked lime, by which means the excess of sulphate and of iron are precipitated in the form of sulphate of lime and protoxyd of iron, from which the clear liquid may be drawn or filtered off.

The bath made, as already described, with 280 pounds of copperas (equal to 56 pounds of iron) and 320 pounds of salt in 2000 pounds of water, has a specific gravity of about 22? Beaumé or 1.150 at ordinary temperatures, water being 1.000. A cubic foot of it weighs 1150 ounces avoirdupois and contains 3.52 lbs. of protochlorid of iron (besides an equal quantity of sulphate of soda), and about $5\frac{1}{2}$ lbs. of salt. This amount of protochlorid contains 1.54 lbs., or 10,780 grains of metallic iron, and as a cubic foot is

---

[1] The reactions between protochlorid of iron and the oxyds of copper are thus expressed in chemical symbols, using, as has been done in the note on page 27, the older notation, in which $Cu = 31.75$, $Fe = 28$, $Cl = 35.5$, and $O = 8$.

For the protoxyd of copper.

$$3Cu_2O_2 + 4FeCl = 2Fe_2O_3 + 2Cu_2Cl + 2CuCl.$$

For the dinoxyd of copper,

$$3Cu_2O + 2FeCl = Fe_2O_3 + 2Cu_2Cl + 2Cu.$$

In the reaction between the protochlorid of copper and the metallic copper,

$$2CuCl + 2Cu = 2Cu_2Cl.$$

equal to almost exactly 1000 fluid-ounces, each fluid-ounce holds in solution 10.78 grains of iron as protochlorid. By the method of assay described above, the amount of iron held in solution by a fluid-ounce of the liquid is very easily determined, and in this way the efficiency of a bath is most conveniently designated. Solutions containing 5.0 grains, and even 3.0 grains of dissolved iron to the fluid-ounce, may be used, but the strongest are most efficient.

The protochlorid of iron serves to chloridize the oxyd of copper in the ore. A cubic foot of bath containing 10,780 grains of dissolved iron will chloridize 18,287 grains (or 2.61 lbs.) of copper in the state of protoxyd, converting one-third of it into protochlorid, and two-thirds of it into dichlorid of copper, of which latter compound (consisting of copper 63.5, chlorine 35.5) there will be formed 29,057 grains, or about 4.15 lbs. The dichlorid is insoluble in water though readily soluble in strong brine, especially if this be heated; hence the necessity of a large excess of salt in the bath. A cubic foot of saturated brine at a temperature of 194° F. will dissolve about 10.0 lbs., and at 104° F. about 5.0 lbs. of the dichlorid, while a cubic foot of brine holding 15 p. c. of salt, will dissolve at 194° F. 6.25 lbs., at 184° F. 3.75 lbs., and at 57° F. 2.18 lbs. of dichlorid of copper, and the same amount of brine holding only 5 p. c. of salt will dissolve at 198° 1.65 lbs., and at 104° 0.70 lbs. of dichlorid. The bath above described, with 5½ pounds of salt to the cubic foot, contains not quite 8 p. c. of salt. It will thus be understood why in some cases it may become necessary to increase the amount of salt in the bath in order to augment its solvent power for the dichlorid. Both by cooling and by dilution with water the dichlorid separates in the form of a white heavy crystalline powder, which is readily converted by simple contact with metallic iron into pure crystalline copper.

In treating copper ores which contain no sulphur and consequently form no soluble sulphate in roasting, the loss of the bath in chlorine may be supplied by adding, from time to time, small portions of protosulphate of iron, or still better by passing over or through the liquid in the stirring or leaching vats, as already described (page 16), a current of sulphurous acid gas. This, being absorbed, converts the separated hydrous peroxyd of iron into a mixture of protosulphate and protosulphite of iron, at the same time liberating the combined chlorine of the oxychlorid in the form of soluble protochlorid.

As the protoxyd of copper is a comparatively feeble base, the solutions of the protochlorid are readily decomposed by the oxyds of zinc and lead, which are often present in roasted ores. These cause the separation from the solutions of a green insoluble oxychlorid composed of oxyd and protochlorid of copper, chlorid of zinc or of lead being formed at the same time.

In the presence of an excess of protochlorid of iron this oxychlorid of copper is immediately dissolved, as oxyd of copper would be, but in the reaction a certain amount of chlorine is consumed in forming the chlorids of zinc and lead. An excess of oxyd of copper also unites with protochlorid of copper to form this oxychlorid, so that in leaching ores charged with oxyd, the protochlorid of copper formed is at first retained in a form insoluble in water and in brine, but as it is completely dissolved in an excess of protochlorid of iron, this reaction gives rise to no difficulty in working.

In like manner carbonate of lime, though without action on solutions of protochlorid of iron below 212° F. readily decomposes protochlorid of copper at 140° F., with separation of a similar oxychlorid, which requires protochlorid of iron to redissolve it. In this way the presence of carbonate of lime in copper ore *indirectly* causes a loss of protochlorid of iron, which must be supplied in one of the ways already set forth. The action of carbonate of magnesia is similar to that of the carbonate of lime. Neither these substances nor the oxyds of lead or zinc separate the copper from the dichlorid.

In the precipitation of the copper by metallic iron 28 parts of this metal unite with 35.5 parts of chlorine, and in so doing separate from a solution of the protochlorid 31.75 of copper, and from the dichlorid twice that amount. Hence to obtain 100 parts of copper from the first requires 88.2 parts, and from the second 44.1 parts of iron, while from solutions in which one-half the copper exists as protochlorid and one-half as dichlorid the amount of iron required will be the mean of these two, or about 66 parts for 100 of copper.

In the roasting of sulphuretted copper ores the greater part of the copper (apart from the sulphate) is obtained as protoxyd, besides a variable amount of dinoxyd, sometimes, according to Plattner, as much as 20 or 30 p. c. of the copper.[1] Such a mixture when treated in the bath gives rise, of course, to a correspondingly large amount of dichlorid, which is, however, generally nearly counterbalanced by the protochlorid resulting from the reaction between the sulphate of copper and the salt of the bath, so that the proportion of iron required to separate 100 parts of copper from the solution of such roasted ores varies from 60 to 70 parts. Hence the present process presents, in this respect, a great economy over the ordinary wet methods in which the precipitation of 100 parts of copper requires 100 and often 120 or more parts of metallic iron.

---

[1] In some cases we have found in such roasted ores a portion of sulphate of dinoxyd of copper. This remains when the ordinary sulphate (of protoxyd) has been removed by water, and may be dissolved from the residue by a hot solution of common salt, by which this insoluble sulphate is converted into dichlorid. Some copper ores of 15 or 20 p. c. have yielded as much as one per cent. of copper in this form, which is of course readily soluble in the protochlorid of iron bath.

A solution of protochlorid of copper when mixed with salt not only has the power of chloridizing and dissolving metallic copper, as already described, but readily takes up the copper from sulphuretted ores, such as the vitreous and variegated species, and from copper matte or regulus, with separation of sulphur and formation of dichlorid. This reaction may, in some cases, be taken advantage of by causing a hot solution nearly saturated with salt and holding protochlorid of copper, to filter through a layer of such ore or regulus in coarse powder. The metal is rapidly taken up, and solutions obtained in which the whole of the copper is present as dichlorid. In this way an additional amount of copper is dissolved and may be separated in the metallic state with very little cost. The 66 parts of iron required to precipitate 100 of copper from the ordinary solutions of mixed protochlorid and dichlorid will separate from solutions of pure dichlorid 150 parts of copper.

V. PATENT SPECIFICATION.

Letters patent for the process above described were granted to T. Sterry Hunt and James Douglas, Jr., in 1869, in the United States, Great Britain and Canada, the date of the United States patent being Feb. 9, 1869. The nature of the process and the mode of applying it having been fully set forth in the preceding pages, it will be sufficient to give the following extracts from the specification:

"We do not claim the use of any particular form of furnace, nor of any special arrangement for calcining, lixiviating or precipitating, reserving to ourselves the choice of the best forms of apparatus for these purposes, neither do we claim the use of protosalts of iron otherwise than in solution, nor the use of perchlorid or other persalts of iron, nor yet the use of sulphurous acid save and except in connection with protosalts of iron, as already set forth.

" What we claim as our invention is:

" I. The use and application of a solution of neutral protochlorid of iron, or of mixtures containing it, for the purpose of converting the oxyd or suboxyd of copper, or their compounds, into chlorids of copper.

" II. The use of sulphurous acid for the purpose of decomposing the oxychlorid of iron formed in the preceding re-action.

" III. The use of a process for the purpose of extracting copper from its naturally or artificially oxydized compounds by the aid of the first, or the first and second of the above reactions, substantially in the manner already set forth."

VI. What Ores of Copper may be Treated by this Process.

The forms in which copper occurs in nature may be conveniently grouped in three classes to each, of which, under certain conditions, the Hunt and Douglas process may be advantageously applied.

In the first class may be included the various sulphuretted ores, such as copper pyrites (often mixed with iron pyrites) and the variegated and vitreous sulphurets, all of which are readily oxydized by calcination. In addition to these are the fahl-ores, which contain besides sulphur, arsenic and antimony. These objectionable elements by calcination are either expelled or rendered insoluble. All the above named ores yield their copper after oxydation to the Hunt and Douglas bath. The question of the comparative fitness of this method for rich and poor ores has already been discussed on page 6.

In the second class are included the oxydized compounds of copper, such as the red and black oxyds, the green and blue carbonates, salts like the oxychlorid and also silicates of copper like chrysocolla. All of these are readily attacked by the bath without previous calcination; but in the case of the red or dinoxyd, as already explained above, it should either be mixed with protoxyd ores or in part converted into protoxyd by a slight calcination in order to render the copper wholly soluble. The carbonates of copper, which are readily dissolved in the bath, give off their carbonic acid so as to cause frothing, to prevent which it may be well to give them a slight calcination or roasting. Heating them to low redness in a kiln or furnace for a few minutes will be sufficient to convert the carbonates into protoxyd.

The common silicate of copper called chrysocolla, readily gives up its copper to the bath of protochlorid of iron and salt, so that its treatment, whether alone or mixed with other ores, presents no difficulty. A peculiar ore which is now treated successfully by this process at Phœnixville, Penn., is a hydrated silicate of oxyd of copper with magnesia, alumina and peroxyd of iron, containing when pure about 13 p. c. of copper. This mineral, which may be described as a copper-chlorite, is readily and completely decomposed by acids but is not attacked by the bath of protochlorid of iron. To extract the copper it is treated as follows: The crude ore, which is mixed with clay and sand and carries from 3.0 to 6.0 p. c. of copper, is heated to low redness for some hours in large vertical muffles each holding 15,000 lbs., having been previously mixed with one-tenth its weight of coal in coarse powder, by which the combined oxyd of copper is reduced to the metallic state. This, on withdrawing the heated charge, is at once oxydized by the air, yielding a mixture of protoxyd and dinoxyd of copper, which are readily and completely removed by the subsequent operation of leaching with the Hunt and Douglas bath. The copper is chiefly dissolved in the form of dichlorid, as is shown by the fact that not more than 50 parts of metallic iron are required to precipitate 100 parts of copper from the solution.

The third class includes the deposits of native or metallic copper, which in almost all instances are most advantageously treated by mechanical means. In those rare cases, in which the copper is too finely divided to be thus profitably extracted, it will be found that by careful calcination at a low red heat it may be oxydized so as to become soluble in the protochlorid of iron bath. In this, as in all other cases of non-sulphuretted ores, it is as already explained (page 16) indispensable to supply the loss of chlorine by the use of sulphurous acid fumes, or by the addition from time to time of a protosalt of iron.

The presence of carbonate of lime or carbonate of magnesia in any ore is objectionable, since as already explained (page 32), it decomposes the protochlorid of copper and thus indirectly precipitates the iron from the bath. The action of oxyds of lead and zinc, which come from the roasting of blende and galena when these are present in the ore, produces a similar effect. When not present in too large quantities, the effect of all these substances may be corrected by careful roasting, which forms a large proportion of sulphates, or by the use of sulphurous fumes, but ores containing much carbonate of lime or carbonate of magnesia are not adapted to treatment by this or any other wet process.

VII. Practical Working of the Process.

The Hunt and Douglas process, after some experimental trials, was first worked continuously for a year in 1872–73 at the Davidson Mine in North Carolina, under the direction of the Messrs. Clayton. The ore, a pyritous copper in a slaty gangue, was dressed up to five or six per cent., crushed to pass through a sieve of forty meshes to the linear inch, roasted in three-hearth reverberatory furnaces so as to contain about one-fourth its copper as sulphate, and treated in stirring vats in charges of 3000 lbs. The loss of copper in the residue was found to be from 0.3 to 0.5 p. c., and the bath maintained its strength in chlorid of iron without the use of copperas or sulphurous acid. The amount of iron consumed was equal to 70 p. c., and the salt, to supply unavoidable losses, to 25 p. c. of the copper produced. These details are from a letter from the manager of the works, Mr. James E. Clayton, published in the *Engineering and Mining Journal* for July, 1873, from which it appears that the entire cost of producing cement copper from the dressed ore of 5½ p. c. was estimated to be three and two-thirds cents a pound.

This mine was subsequently abandoned, and the same proprietors in 1874 erected works with six calcining furnaces for the treatment of twelve tons of pyritous ore daily by this process at the Ore Knob mine in Ashe County, North Carolina. Up to the first of January, 1875, over 200 tons of copper had there been made by this process. In the report bearing that date of the directors of the Ore Knob Co., James E. Tyson of Baltimore, president, it is said, "From the data furnished by the superintendent in his Report from the mine, and a careful estimate made here, we find the cost of making copper, mining, and all expenses included, to be less than eight cents a pound."

These works were soon after enlarged to nearly three times their former capacity, but in sinking below the water-line in the mine the ore, hitherto free from lime, was found to contain 30 p. c. or more of carbonate of lime with some magnesia. The direct treatment of such an ore by any moist process was impracticable, and the reduction works were accordingly suspended pending the erection of dressing-works in which it is proposed to concentrate the ore by crushing and washing, removing thereby the carbonate of lime of the gangue. The concentrating machinery, as we are informed by the managing director of the Ore Knob Copper Co., Mr. James E. Clayton, will be in operation in June, 1876, when it is proposed to recommence at once the treatment of the purified ores by the Hunt and Douglas process.

Reduction-works are now in successful operation at Phœnixville, Pennsylvania, where copper ores of two kinds are treated by the Hunt and Douglas process, the first of which is a magnetic iron ore from Berks Co., Penn., containing about 3 p. c. of copper, chiefly as copper pyrites, mixed, however, with a little carbonate and silicate of copper. This ore, of which 20,000 lbs. are treated daily, is crushed so as to pass through a sieve of seven meshes to the linear inch, roasted as already explained on page 11, and subsequently treated by leaching. The residue, which contains about 0.5 p. c. of copper, is a rich iron ore which is used for lining puddling furnaces. The second ore is the peculiar hydrated silicate described on page 34, of which 15,000 lbs. are treated daily. The leached residues of this do not retain over 0.3 p. c. of copper.

The works of the Stewart Reduction Co., at Georgetown, Colorado, in which this process is applied to mixtures of silver and copper ores, will be again referred to.

VIII. Treatment of Silver and Gold Ores.

The use of soluble compounds of copper as an agent in treating silver ores and rendering them fit for amalgamation, has long been known, and is the basis of the Mexican patio process and its modifications, as well as of the Washoe process now largely employed in the west. The theory of the action of the copper salts in the first of these methods, where the materials are exposed for a long time to the action of the air, is still somewhat obscure. In the Washoe method sulphate of copper and common salt are added together to the ground ore mixed with water, and from these by the reactions which take place in the pans, dichlorid of copper is soon formed. This substance dissolved in brine is used directly with advantage in the treatment of silver ores by Janin and by Kröncke. From the results of various experimenters, it is clear that solutions, both of protochlorid and dichlorid of copper, mixed with common salt, when at an elevated temperature, effect a complete chlorination of sulphuretted and arsenical silver ores, or at least render them susceptible of ready and complete amalgamation.

The use of the chlorids of copper as hitherto applied, presents, however, several difficulties : 1st. The sulphate of copper from which they are generally prepared is costly, and in some places difficult to procure; 2d. Protochlorid of copper is readily decomposed and separated from hot solutions as an insoluble oxychlorid by the carbonate of lime often found with the ores; 3d. Solutions of dichlorid of copper in brine very readily absorb oxygen from the air, forming, besides protochlorid of copper, also an insoluble oxychlorid. These oxychlorids are without action on silver ores, though they attack the mercury when amalgamation is attempted simultaneously with the treatment by copper salts, forming an insoluble chlorid of this metal, and thereby causing a considerable loss.

To meet these objections there is needed a cheap and ready method of preparing the chlorids of copper, and a simple means of preventing their precipitation in inert or noxious forms by the action of the air or carbonate of lime. It will be apparent from the preceding account of the chemistry of the Hunt and Douglas copper process, that the use of a heated solution of protochlorid of iron and salt, aided by sulphurous acid, for the solution of the oxydized compounds of copper, meets the conditions of the problem in the following manner:

1st. The Hunt and Douglas bath gives readily and cheaply strong solutions of the mixed protochlorid and dichlorid of copper wherever carbonates, oxyds, or calcined sulphuretted ore of this metal can be had.

2d. It dissolves the oxychlorids of copper, by whatever means produced, changing them into a mixture of protochlorid and dichlorid of copper, and thus prevents any deterioration of the copper solution by the action of the air or of carbonate of lime.

The Hunt and Douglas bath may be advantageously applied:[1]

I. To effect more cheaply and more completely the chlorination and the amalgamation of such silver ores as are now treated in the raw state with chemicals, as they are called, — that is to say, sulphate or chlorid of copper with common salt.

II. To chlorinate such silver ores as have been calcined without the addition of salt.

III. To complete the chlorination of silver ores which have been partially chlorinated by calcining with salt, thus securing a much more complete extraction of the silver than has hitherto been attained.

In all of these cases it will be understood that some oxydized form of copper, such as carbonate, native oxyd or calcined sulphuretted ore, is to be added, unless it is already present in the silver ore to be treated. It may be added even in large quantities with advantage, and from the solutions charged with copper a portion, or the whole of this metal may be precipitated from time to time by metallic iron as cement copper.

In localities where salts of iron are not readily obtained, and where sulphur ores are abundant, it will be found that by passing sulphurous acid gas into or over a solution of salt holding pulverized oxyd or carbonate of copper in suspension, a solution of dichlorid of copper will be readily formed, and this reaction may be rendered available for the treatment of silver ores. By precipitating the copper solution thus obtained with metallic iron, protochlorid of iron is at once readily and cheaply obtained.

Silver ores chlorinated by the Hunt and Douglas bath, may be subsequently treated, either by dissolving the silver from the washed residues by a solution of hyposulphite or of chlorid of sodium, or by amalgamation. The use of mercury is to be preferred for ores holding, besides silver, a portion of gold. Such ores should be treated with the bath in the raw state, or after simple calcination, roasting with salt being for them objectionable.

United States letters patent (No. 151,763) for the use of the Hunt and Douglas bath of protochlorid of iron and common salt, conjointly with sulphurous acid, for the treatment of silver ores, or silver and gold ores, mixed with oxydized ores of copper, were granted June 9, 1874, to James Douglas, Jr., Thomas Sterry Hunt and James Oscar Stewart. This process has now been most successfully applied for more than a year on a large scale in the working of silver ores by Mr. Stewart, who will publish in the

[1] Later observations show that this process may be advantageously applied to the treatment of the tellurids of silver and gold.

course of the summer of 1876 a detailed description of the method as adapted by him to various kinds of silver ores. Copies of this (and also of the present pamphlet) may be had by addressing J. Oscar Stewart, Georgetown, Colorado.

For further information concerning the Hunt and Douglas process as applied to copper extraction, address

JAMES DOUGLAS, Jr.,

Phœnixville, Penn.,

or

T. STERRY HUNT,

Boston, Mass.

May, 1876.

www.ingramcontent.com/pod-product-compliance
Lightning Source LLC
Chambersburg PA
CBHW022031190326
41519CB00010B/1670